THE DOCTRINE OF NATURAL SELECTION

BY

ALFRED RUSSEL WALLACE

British Library Cataloguing-in-Publication Data
A catalogue record for this book is available from the
British Library

Alfred Russel Wallace

Alfred Russel Wallace was born on 8th January 1823 in the village of Llanbadoc, in Monmouthshire, Wales.

At the age of five, Wallace's family moved to Hertford where he later enrolled at Hertford Grammar School. He was educated there until financial difficulties forced his family to withdraw him in 1836. He then boarded with his older brother John before becoming an apprentice to his eldest brother, William, a surveyor. He worked for William for six years until the business declined due to difficult economic conditions.

After a brief period of unemployment, he was hired as a master at the Collegiate School in Leicester to teach drawing, map-making, and surveying. During this time he met the entomologist Henry Bates who inspired Wallace to begin collecting insects. He and bates continued exchanging letters after Wallace left teaching to pursue his surveying career. They corresponded on prominent works of the time such as Charles Darwin's *The Voyage of the Beagle* (1839) and Robert Chamber's *Vestiges of the Natural History of Creation* (1844).

Wallace was inspired by the travelling naturalists of the day and decided to begin his exploration career collecting specimens in the Amazon rainforest. He explored the Rio

Negra for four years, making notes on the peoples and languages he encountered as well as the geography, flora, and fauna. On his return voyage his ship, Helen, caught fire and he and the crew were stranded for ten days before being picked up by the Jordeson, a brig travelling from Cuba to London. All of his specimens aboard Helen had been lost.

After a brief stay in England he embarked on a journey to the Malay Archipelago (now Singapore, Malaysia, and Indonesia). During this eight year period he collected more than 126,000 specimens, several thousand of which represented new species to science. While travelling, Wallace refined his thoughts about evolution and in 1858 he outlined his theory of natural selection in an article he sent to Charles Darwin. This was published in the same year along with Darwin's own theory. Wallace eventually published an account of his travels *The Malay Archipelago* in 1869, and it became one of the most popular books of scientific exploration in the 19th century.

Upon his return to England, in 1862, Wallace became a staunch defender of Darwin's landmark work *On the Origin of Species* (1859). He wrote responses to those critical of the theory of natural selection, including 'Remarks on the Rev. S. Haughton's Paper on the Bee's Cell, And on the Origin of Species' (1863) and 'Creation by Law' (1867). The former of these was particularly pleasing to Darwin. Wallace also published important papers such as 'The Origin of Human

Races and the Antiquity of Man Deduced from the Theory of 'Natural Selection" (1864) and books, including the much cited *Darwinism* (1889).

Wallace made a huge contribution to the natural sciences and he will continue to be remembered as one of the key figures in the development of evolutionary theory.

Wallace died on 7th November 1913 at the age of 90. He is buried in a small cemetery at Broadstone, Dorset, England.

THE DOCTRINE OF NATURAL SELECTION

NOTWITHSTANDING the objections which are still made to the theory of Natural Selection, on the ground that it is either a pure hypothesis not founded on any demonstrable facts, or a mere truism which can lead to no useful results, we find it year by year sinking deeper into the minds of thinking men, and applied, more and more frequently, to elucidate problems of the highest importance. In the works now before us we have this application made by two eminent writers, one a politician, the other a naturalist, as a means of working out so much of the complex problem of human progress as more especially interests them.

Mr. Bagehot takes for granted that early progress of man which resulted in his separation into strongly-marked races, in his acquisition of language, and of the rudiments of those moral and intellectual faculties which all men possess; and his object is to work out the steps by which he advanced to the condition in which the dawn of history finds him— aggregated into distinct societies known as tribes or nations, subject to various forms of government, influenced by various beliefs and prejudices, and the slave of habits and customs which often seem to us not only absurd and useless, but even positively injurious. Now, every one of these beliefs or customs, or these aggregations of men into groups having some common characteristics, must have been useful at the time they originated; and a great feature of Mr. Bagehot's little book is his showing how even the most unpromising of these, as we now regard them, might have been a positive step in advance when they first appeared. His main idea is, that what was wanted in those early times was some means of combining men in societies, whether by the action of some common belief or common danger, or by the power of some ruler or tyrant. The mere fact of obedience to a ruler was at first much more important than what was done by means of the obedience. So, any superstition or any custom, even if it originated in the grossest delusion, and produced positively bad results, might yet, by forming a bond of union more perfect than any other then existing, give the primitive tribe subject to it such a relative advantage over the disconnected families around them as to lead to their increase and permanent survival in the struggle for existence. In those early days war was perhaps the most powerful means of forcing men to combined action, and might therefore have been necessary for the ultimate development of civilization. Freedom of opinion was then a positive evil, for it would lead to independent action, the very thing it was most essential to get rid of. In early times isolation was an advantage, in order that these incipient societies might not be broken up by intermixture, and it was only after a large number of such little groups, each with its own idiosyncrasies, habits, and beliefs, had been formed, that it became advantageous for them to meet to intermingle or to struggle together, and the stronger to drive out or exterminate the weaker. Out of the great number of petty tribes thus formed, only a few had the qualities which led to a further advancement. The rest were either exterminated or driven out into remote and inaccessible or inhospitable districts, and some of those are the "savages" which still exist on the earth, serving as a measure of the vast progress of the human race. Yet even these never show us the condition of the primitive man; they are men who advanced up to a certain point and then became stationary:

"Their progress was arrested at various points; but nowhere, not even in the hill-tribes of India, not even in the Andaman Islands, not even in the savages of Terra del Fuego, do we find men who have not got some way. They have made their little progress in a hundred different ways; they have framed with infinite assiduity a hundred curious habits; they have, so to say, *screwed* themselves into the uncomfortable corners of a complex life, which is odd and dreary, but yet is possible. And the corners are never the same in any two parts of the world. Our record begins with a thousand unchanging edifices, but it shows traces of previous building. In historic times there has been but little progress, in prehistoric times there must have been much."

Again our author shows how valuable must have been

the institution of caste in a certain stage of progress. It established the division of labor, led to great perfection in many arts, and rendered government easy. Caste nations would at first have a great advantage over non-caste nations, would conquer them, and increase at their expense. But a caste nation at last becomes stationary; for a habit of action and a type of mind which it can with difficulty get rid of are established in each caste. When this is the case, non-caste nations soon catch them up, and rapidly leave them far behind.

This outline will give some idea of the way in which Mr. Bagehot discusses an immense variety of topics connected with the progress of societies and nations, and the development of their distinctive peculiarities. The book is somewhat discursive and sketchy, and it contains many statements and ideas of doubtful accuracy, but it shows an abundance of ingenious and original thought. Many will demur to the view that mere accident and imitation have been the origin of marked national peculiarities; such as those which distinguish the German, Irish, French, English, and Yankees: "The accident of some predominant person possessing certain peculiarities set the fashion, and it has been imitated to this day." And again: "Great models for good or evil sometimes appear among men who follow, them either to improvement or degradation." This is said to be one of the chief agents in "nation-making," but a much better one seems to be the affinity of like for like, which brings and keeps together those of like morals, or religion, or social habits; but both are probably far inferior to the long-continued action of external Nature on the organism, not merely as it acts in the country now inhabited by the particular nation, but by its action during remote ages and throughout all the migrations and intermixtures that our ancestors have ever undergone. We also find many broad statements as to the low state of morality and of intellect in all prehistoric men, which facts hardly warrant, but this is too wide a question to be entered upon here. In the concluding chapter, "The Age of Discussion," there are some excellent remarks on the restlessness and desire for immediate action which civilized men inherit from their savage ancestors, and how much it has hindered true progress; and the following passage, with which we will conclude the notice of Mr. Bagehot's book, might do much good if, by means of any skilful surgical operation, it could be firmly fixed in the minds of our legislators and of the public:

"If it had not been for quiet people, who sat still and studied the sections of the cone; if other people had not sat still and worked out the doctrine of chances, the most 'dreamy moonshine,' as the purely practical mind would consider, of all human pursuits; if 'idle star-gazers' had not watched long and carefully the motions of the heavenly bodies—our modern astronomy would have been impossible; and, without astronomy, 'our ships, our colonies, our seamen,' all that makes modern life, could not. have existed. Ages of quiet, sedentary, thinking people were required before that noisy existence began, and without those pale, preliminary students it never could have been brought into being. And nine-tenths of modern science is, in this respect, the same; it is the produce of men whom their contemporaries thought—dreamers who were laughed at for caring for what did not concern them—who, as the proverb went, 'walked into a well from looking at the stars'—who were believed to be useless, if any one could be such. And the conclusion is plain that, if there had been more such people; if the world had not laughed at those there were; if, rather, it had encouraged them—there would have been a great accumulation of proved science ages before there was. It was the irritable activity, 'the wish to be doing something,' that prevented it. Most men inherited a nature too eager and too restless to find out things; and, even worse—with their idle clamor they disturbed the brooding hen, 1 they would not let those be quiet who wished to be so, and out of whose calm thought much good might have come forth. If we consider how much good science has done, and how much it is doing for mankind, and, if the over-activity of men is proved to be the cause why science came so late into the world, and is so small and scanty still, that will convince most people that our over-activity is a very great evil."

In the second work, of which we have given the title, the veteran botanist, Alphonse de Candolle, sets forth his ideas on many subjects not immediately connected with the science in which he is so great an authority. The most important, though not the longest, essay in the volume is that on "Selection in the Human Race," in which he arrives at some results which differ considerably from those of previous writers. In a section on "Selection in Human Societies or Nations," we find a somewhat novel generalization as to the progress and decay of nations. Beginning with small, independent states, we see a gradual fusion of these into larger and larger nations, sometimes voluntary, sometimes by conquest, but the fusion always goes on, and tends to become more and more complete, till we have enormous aggregations of people under one government, in which local institutions gradually disappear, and result in an almost complete political and social uniformity. Then commences decay; for the individual is so small a unit, and so powerless to influence the government, that the mass of men resign themselves to passive obedience. There is then no longer any force to resist internal or external enemies, and by means of one or the other the "vast fabric" is dismembered, or falls in ruins. The Roman Empire and the Spanish possessions in America are examples of this process in the past; the Russian Empire and our Indian possessions will inevitably follow the same order of events in a not very distant future.

Although M. de Candolle is a firm believer in Natural Selection, he takes great pains to show how very irregular and uncertain it is in its effects. The constant struggles and wars among savages, for example, might be supposed to lead to so rigid a selection that all would be nearly equally strong and powerful; and the fact that some savages are so weak and incapable as they are shows, he thinks, that the action of natural selection has been checked by various incidental causes. He omits to notice, however, that the struggle between man and the lower animals was at first the severest, and probably had a considerable influence in determining race-characters. It may be something more than accidental coincidence that the most powerful of all savages—the negroes—inhabit a country where dangerous wild beasts most abound; while the weakest of all—the Australians—do not come into contact with a single wild animal of which they need be afraid.

Selection among barbarous nations will often favor cunning, lying, and baseness; vice will gain the advantage, and nothing good will be selected but physical beauty. Civilization is defined by the preponderance of three facts— the restriction of the use of force to legitimate defence and the repression of illegitimate violence, speciality of professions and of functions, and individual liberty of opinion and action under the general restriction of not injuring others. By the application of the above tests we can determine the comparative civilization of nations; but too much civilization is often a great danger, for it inevitably leads to such a softening of manners, such a hatred of bloodshed, cruelty, and injustice, as to expose a nation to conquest by its more warlike and less scrupulous neighbors. Progress in civilization must necessarily be very slow, and to be permanent must pervade all classes and all the surrounding nations; and it is because past civilizations have been too partial that there have been so many relapses into comparative barbarism. All this is carefully worked out, and is well worthy of attention.

In the last section, on the probable future of the human race, we have some remarkable speculations, very different from the somewhat Utopian views held by most evolutionists, but founded, nevertheless, on certain very practical considerations. In the next few hundred or a thousand years the chief alterations will be the extinction of all the less dominant races, and the partition of the world among the three great persistent types, the whites, blacks, and Chinese, each of which will have occupied those portions of the globe for which they are best adapted. But, taking a more extended glance into the future of 50,000 or 100,000 years hence, and supposing that no cosmical changes occur to destroy, wholly or partially, the human race, there are certain well-ascertained facts on which to found a notion of what must by that time have occurred. In the first place, all the coal and all the metals available will then have been exhausted, and, even if men succeed in finding other sources of heat, and are able to extract the metals thinly diffused through the soil, yet these products must become far dearer and less available for general use than now. Railroads and steam-ships, and every thing that depends upon the possession of large quantities of cheap metals, will then be impossible, and sedentary agricultural populations in warm and fertile regions will be the best off. Population will have lingered longest around the greatest masses of coal and iron, but will finally become most densely aggregated within the tropics. But another and more serious change is going on, which will result in the gradual diminution and deterioration of the terrestrial surface. Assuming the undoubted fact that all our existing land is wearing away and being carried into the sea, but, by a strange over-sight, leaving out altogether the counteracting internal forces, which for countless ages past seem always to have raised ample tracts above the sea as fast as subaërial denudation has lowered them, it is argued that, even if all the land does not disappear and so man become finally extinct, yet the land will become less varied, and will consist chiefly of a few flat and parched-up plains, and volcanic or coralline islands. Population will by this time necessarily have much diminished, but it is thought that an intelligent and persevering race may even then prosper. "They will enjoy the happiness which results from a peaceable existence, for, without metals or combustibles, it will be difficult to form fleets to rule the seas, or great armies to ravage the land;" and the conclusion is that "such are the probabilities according to the actual course of things." Now, although we cannot admit this to be a probability on the grounds stated by M. de Candolle, it does seem a probability, at some more distant epoch, on other grounds. The great depths of the oceans extend over wide areas, whereas the great heights of the land are only narrow ridges and peaks; hence it has been calculated that the mean height of the land is only 1,000, while the mean depth of the sea is about 15,000 feet. But the sea is $2\frac{1}{2}$ times as extensive as the land, so that the bulk or mass of the land above the sea-level will be only about one thirty-seventh of the mass of the ocean. Now, does not this small proportion of bulk of land to water render it highly probable that the forces of elevation and depression should sometimes cause the total or almost total submersion of the land? Of such an epoch no geological record could be left because there could be no strata formed, except from the *débris* of coral-islands, and such a period of destruction of the greater part of terrestrial life may have repeatedly occurred between the period when the several Primary or Secondary formations were deposited. At all events, with such a proportion of land and sea surface as now exists, with such a small bulk of land above the enormous bulk of water, and with no known cause why the dry land rather than the sea-bottom should be constantly elevated, we must admit it to be almost certain that great fluctuations of the area of the land must occur, and that, while those fluctuations could not very considerably increase the area of the land they might immensely diminish it. There is here, therefore, a cause for the possible depopulation of the earth likely to occur much sooner than any cosmical catastrophe.

The largest and most elaborate essay in the volume is that on the "History of the Sciences and of Scientific Men for the last Two Centuries." In this the author endeavors to arrive at certain conclusions as to the progress of science under different conditions and in different countries, the influence of political institutions and of heredity, and various other phenomena, by a method which is novel and ingenious. He takes account only of the men honored as foreign associates or members by the three great European scientific bodies, the Royal Society of London and the Paris and Berlin Academies. By this means he avoids all personal bias, and secures, on the whole, impartiality. The tables drawn out by this method are examined in every possible way, and the results worked out in the greatest detail. The main conclusion arrived at is the determination of a series of eighteen causes favorable to the progress of science; and it is shown that a large proportion of these are present in a considerable degree in countries where science flourishes, while they are almost wholly absent in barbarous or semi-civilized countries where science does not exist.

Another interesting essay is that on the importance for science of a dominant language, and it contains some very curious facts as to the way in which the English language is spreading on the Continent. M. de Candolle believes that in less than two centuries English will be the dominant language, and will be almost exclusively used in scientific works.

There are also short but very interesting essays on methods of teaching drawing and developing the observing powers of children, on statistics and free-will, and on a few other subjects of less importance, all of which are treated in a thoughtful manner, and illustrate one of the views on which much stress is laid in this work, viz., that the mental faculties which render a man great in any science are not special, but would enable him to attain equal eminence in many other branches of science or in any professional or political career.—*Nature.*

↑ "Physics and Politics; or, Thoughts on the Application of the Principles of 'Natural Selection' and 'Inheritance' to Political Society." By Walter Bagehot. (King & Co., 1872.) "Histoire des Sciences et des Savants depuis deux Siècles, suivie d'autres Études sur les Sujets Scientifiques, en particulier sur la Sélection dans l'Espèce Humaine." Par Alphonse de Candolle. (Genève: H. Georg, 1873.)